THE FACES, ERR PHASES, OF THE MOON

ASTRONOMY BOOK FOR KIDS REVISED EDITION | CHILDREN'S ASTRONOMY BOOKS

BABY PROFESSOR

Revised Edition, 2019

Published in the United States by Speedy Publishing LLC, 40 E Main Street, Newark, Delaware 19711 USA.

© 2019 Baby Professor Books, an imprint of Speedy Publishing LLC

All rights reserved.

Without limiting the rights under the copyright reserved above, no part of this publication may be reproduced, stored in or introduced into a retrieval system, or transmitted, in any form, or by any means (electronic, mechanical, photocopying, recording, or otherwise), without the prior written permission of the copyright owner.

All images in this book have been reproduced with the knowledge and prior consent of the artists concerned, and no responsibility is accepted by producer, publisher, or printer for any infringement of copyright or otherwise arising from the contents of this publication.

Baby Professor Books are available at special discounts when purchased in bulk for industrial and sales-promotional use. For details contact our Special Sales Team at Speedy Publishing LLC, 40 E Main Street, Newark, Delaware 19711 USA. Telephone (888) 248-4521 Fax: (210) 519-4043. www.speedybookstore.com

10 9 8 7 6 * 5 4 3 2 1

Print Edition: 9781541968257
Digital Edition: 9781541968493
Hardcover Edition: 9781541968394

See the world in pictures. Build your knowledge in style.
https://www.speedypublishing.com/

CONTENTS

UNDERSTANDING THE PHASES OF THE MOON 10

THE MOON'S LIGHT COMES FROM THE SUN 14

THE PHASES OF THE MOON 18

WHAT IS WAXING AND WHAT IS WANING? 22

WHAT'S THE DIFFERENCE BETWEEN
A DARK MOON AND A NEW MOON? 24

THE EIGHT PHASES OF THE MOON IN ORDER 26

DIFFERENCE BETWEEN NORTHERN
AND SOUTHERN HEMISPHERES 44

WHAT IS A BLUE MOON? 46

THE DARK SIDE OF THE MOON 48

MOON PHASES
CHART

In this book, we're going to talk about the eight different phases of Earth's moon. So, let's get right to it!

Have you ever wondered why the moon looks different every night? Sometimes it looks like a big, round shiny ball. Sometimes it looks like a banana and sometimes it doesn't show up in the sky at all. Ancient peoples wondered why it disappeared completely and why it came back. Of course, today we know that it hasn't disappeared, but you might wonder why it goes through a regular pattern of different "faces" every month.

A full moon looks like a big, round shiny ball.

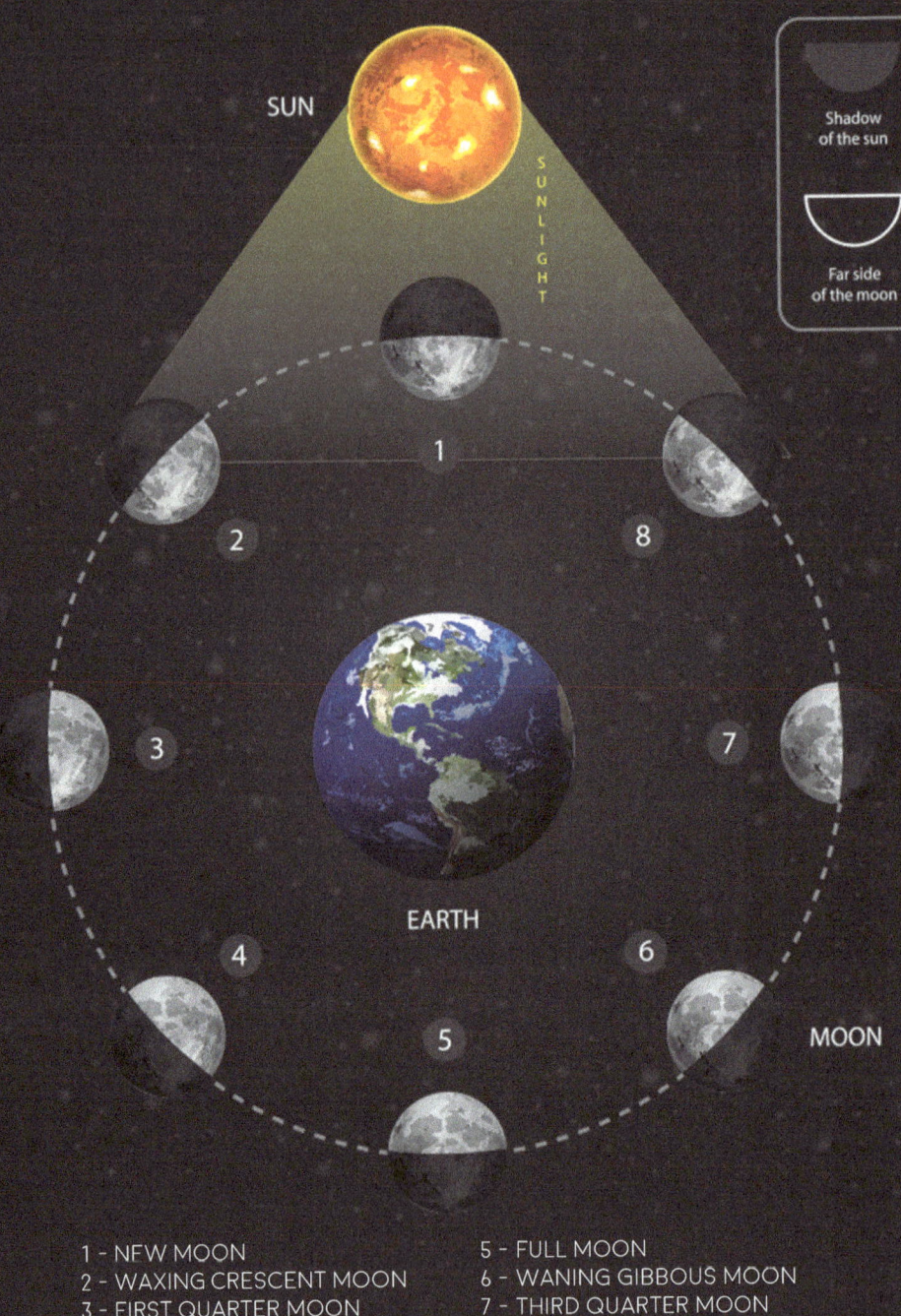

Some people think that it's Earth's shadow that causes the different looks of the moon, but this isn't true. The different "faces" that we see of the moon are called its phases. There are eight phases it goes through every 29.5305882 days or approximately every 29.5 days. The phases of the moon are the basis for our monthly calendar.

UNDERSTANDING THE PHASES OF THE MOON

The moon is Earth's only natural satellite. There are manmade satellites that orbit the Earth too. As the moon goes around planet Earth, its position changes relative to the sun and this makes the moon go through its regular phases. From our Earth viewpoint, the moon changes its shape every night but, of course, the physical shape of the moon doesn't change. It's only the pattern of reflected sunlight that changes.

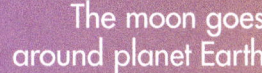

The moon goes around planet Earth

Primitive peoples were intrigued by these changes and used the shape-shifting moon to create a lunar calendar. The Muslim as well as the Hebrew and Chinese calendars are all set up this way. The New Moon is at the beginning of the calendar month and Full Moon always falls mid-month. However, most modern calendars don't work this way because the year would end up being 12 days too short. A lunar month at 29.53 days is actually a little shorter than the standard month that we use in our calendar today, which is an average of 30.44 days.

LUNAR · CALENDAR

JANUARY
S	M	T	W	T	F	S
	1	2	3	4	5	
6	7	8	9	10	11	12
13	14	15	16	17	18	19
20	21	22	23	24	25	26
27	28	29	30	31		

FEBRUARY
S	M	T	W	T	F	S
					1	2
3	4	5	6	7	8	9
10	11	12	13	14	15	16
17	18	19	20	21	22	23
24	25	26	27	28		

MARCH
S	M	T	W	T	F	S
					1	2
3	4	5	6	7	8	9
10	11	12	13	14	15	16
17	18	19	20	21	22	23
24	25	26	27	28	29	30
31						

APRIL
S	M	T	W	T	F	S
	1	2	3	4	5	6
7	8	9	10	11	12	13
14	15	16	17	18	19	20
21	22	23	24	25	26	27
28	29	30				

MAY
S	M	T	W	T	F	S
			1	2	3	4
5	6	7	8	9	10	11
12	13	14	15	16	17	18
19	20	21	22	23	24	25
26	27	28	29	30	31	

JUNE
S	M	T	W	T	F	S
						1
2	3	4	5	6	7	8
9	10	11	12	13	14	15
16	17	18	19	20	21	22
23	24	25	26	27	28	29
30						

JULY
S	M	T	W	T	F	S
	1	2	3	4	5	6
7	8	9	10	11	12	13
14	15	16	17	18	19	20
21	22	23	24	25	26	27
28	29	30	31			

AUGUST
S	M	T	W	T	F	S
				1	2	3
4	5	6	7	8	9	10
11	12	13	14	15	16	17
18	19	20	21	22	23	24
25	26	27	28	29	30	31

SEPTEMBER
S	M	T	W	T	F	S
1	2	3	4	5	6	7
8	9	10	11	12	13	14
15	16	17	18	19	20	21
22	23	24	25	26	27	28
29	30					

OCTOBER
S	M	T	W	T	F	S
		1	2	3	4	5
6	7	8	9	10	11	12
13	14	15	16	17	18	19
20	21	22	23	24	25	26
27	28	29	30	31		

NOVEMBER
S	M	T	W	T	F	S
					1	2
3	4	5	6	7	8	9
10	11	12	13	14	15	16
17	18	19	20	21	22	23
24	25	26	27	28	29	30

DECEMBER
S	M	T	W	T	F	S
1	2	3	4	5	6	7
8	9	10	11	12	13	14
15	16	17	18	19	20	21
22	23	24	25	26	27	28
29	30	31				

NORTHERN HEMISPHERE

THE MOON'S LIGHT COMES FROM THE SUN

The moon doesn't have its own light source. The shine that we see when we look at the moon is actually reflected sunlight. You can do a very simple experiment yourself to see how this works. Take a lampshade off a lamp and set it up in a very dark room. The darker the room is, the better your experiment will work. The lamp will represent the sun. Get a smooth, white foam ball about 4 inches wide and stick a long pencil through the bottom of it. This ball will represent the moon. Your head is going to be the Earth as you're observing from the Earth.

Only one side of the moon faces the Earth. We never see its dark side because it's tidally locked with us. For the purposes of this experiment, all you have to do is keep the sun in the same place and have the moon revolve around you—the Earth. As the moon moves, you'll notice that one side of it is always lit by the sun. However, depending on where the moon is located in its 29.5 day orbit, we on Earth only see part of that sunlit area. The rest is in shadow. As the moon gradually orbits the Earth, the amount we see of the sunlight reflection waxes and wanes, which simply means it increases and then lessens.

Only one side of the moon faces the Earth.

THE PHASES OF THE MOON

The phases change as the moon goes around the Earth. During the New Moon and Dark Moon phases, we can't see any reflection of sunlight from the moon's surface. At this point in its orbit, the moon is located between the Earth and the sun.

New Moon - Alignment of Sun, Moon and Earth

Sun

Moon

Earth

As the moon continues on its path around the Earth, an observer can see more and more of the side that is lit by the sun. Finally, the moon gets to the Earth's opposite side. At this point, we can see a bright, shiny full moon. As the moon continues its travels around the Earth, we are seeing less of the sunlight with each passing day until we can't see even a sliver of reflected sunlight at all.

An observer can see more and more of the side that is lit by the sun.

WHAT IS WAXING AND WHAT IS WANING?

As the moon continues from when we don't see it at all to when we see it as a full moon, the light that we see continues to increase every night. This period of the moon's cycle is called waxing. After we see the full moon, the light starts to decrease every night until it gets to the dark moon phase. This decrease in light is called waning.

full moon

waning moon

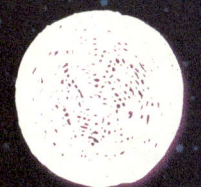  

waxing moon

new moon

WHAT'S THE DIFFERENCE BETWEEN A DARK MOON AND A NEW MOON?

The dark moon is the time period when an observer from Earth can't see the moon in the sky. This time lasts between 1-3 days depending on where you are observing from Earth. It's the time range between when the waning or decreasing Moon ends and the waxing or increasing Moon begins again.

The new moon occurs at exactly the midpoint of the dark moon phase. At this point, the moon is right in between the decreasing and increasing phases.

THE EIGHT PHASES OF THE MOON IN ORDER

The eight phases of the moon in order are:

(1)—the new moon phase;
(2)—the waxing crescent;
(3)—the first quarter moon phase,
(4)—the waxing gibbous moon;
(5)—the full moon,
(6)—the waning gibbous moon;
(7)—the last quarter moon; and
(8)—the waning crescent moon.

Then, the cycle starts over again. The moon gradually moves from phase to phase and looks a little different every night. Let's look at each phase in detail in the order in which they cycle through every month.

When the moon is positioned directly between the Earth's position and the sun's position, we can't see any light reflected from its surface. This is when it goes into the "dark moon" phase, which lasts 1-3 days. The midpoint of this cycle is called the "new moon." A solar eclipse can only happen when the moon is in the new moon phase of its cycle.

New Moon

Waxing Crescent Moon

This phase is a crescent shape that looks something like a banana with its curved edge to the right. A waxing crescent means that on subsequent days, the observer will see more and more of the sunlit portion of the moon.

This phase is called the "first quarter" because by this time the moon has made one-quarter or one-fourth of its trip around the Earth. It's somewhat confusing through because you as an observer would see the right-hand side of the moon lit up and the left-hand side in shadow, half and half.

First Quarter Moon

During this phase, the sunlight on the moon's face has gotten larger and is spreading over to the left side. We're still in the waxing stage here, which means that the amount of sunlight we see reflected night after night is increasing.

Waxing Gibbous Moon

When the moon is positioned directly opposite to its position at new moon, then an Earth observer sees a bright, round, full shiny moon. This occurs when instead of being between the Earth and the sun it is on Earth's opposite side. This position is called opposition. This is the only phase position where a lunar eclipse can occur.

Full Moon

Now that we've reached the full moon, there's only one place to go and that's the beginning of the decreasing phase. As it starts to decrease, sunlight disappears from the right edge of the moon and keeps moving left.

Waning Gibbous Moon

At this point, the last one-fourth of the moon's complete cycle around the Earth, the sunlight covers over the left side of the moon's "face" and the right-hand side is in darkness.

Last Quarter Moon

The waning crescent once again looks banana-shaped with its curved edge facing the left this time. The sunlight decreases with every passing day until the dark moon occurs and we can't see the moon in the sky for a few days.

Waning Crescent Moon

DIFFERENCE BETWEEN NORTHERN AND SOUTHERN HEMISPHERES

The descriptions of the moon's phases in this book are only for the northern hemisphere. For the southern hemisphere, the viewpoints are flipped. For example, for the Last Quarter Moon, in Australia this moon would appear with sunlight over its right side and shadow over its left.

WHAT IS A BLUE MOON?

> Usually during a calendar month we only have one full moon.

sually during a calendar month we only have one full moon. But sometimes, since the modern calendar is not a lunar calendar, there's a second full moon within the same month. This only happens once every three years or so. This event is where the expression "once in a blue moon" comes from.

THE DARK SIDE OF THE MOON

All these phases of the moon are on the same side of the moon, the only side we see. That side is always pointed toward us. The other side is called the "dark side" because it never faces us, which means we are tidally locked.

Earth behind the far side of the Moon

Awesome! Now you know more about why we see different phases of the moon. You can find more Astronomy books from Baby Professor by searching the website of your favorite book retailer.

Visit
www.SpeedyBookStore.com

To view and download free content on your favorite subject and browse our catalog of new and exciting books for readers of all ages.

CPSIA information can be obtained
at www.ICGtesting.com
Printed in the USA
LVHW081246111220
673930LV00017B/80